AI and the Black Church

Bridging Faith, Justice and Innovation

"What Was Meant for Evil, God Meant for Good."

Genesis 50:20

Reverend Dr. Sineta Scott Robertson

AI and the Black Church:

Bridging Faith, Justice and Innovation

Published by: Perspectives for Peace, LLC

ISBN: 978-1-7367968-2-5

Printed in the United States

Dedication

This book is dedicated to Reverend Dr. John L. McCoy and Reverend Dr. Clevester O. Wimbish, President and Vice President respectively of the Washington Baptist Theological Seminary in Mount Rainier, Maryland. Their leadership is steady and resolute, and they have consistently challenged students to dig deep into Scripture, uncover truth, and faithfully apply it to their lives today. I am grateful for the way they model what it looks like to lead with wisdom and conviction.

This book is also dedicated to the pastors and church leaders who are earnestly seeking to bridge the Word of God with the reality of Artificial Intelligence. Your honest questions and thoughtful concerns about AI inspired me to search the Scriptures more diligently, to kneel in prayer, and to ask how God would have His people respond to this emerging technology. You are the reason this book exists. I pray it gives you tools not just for your minds, but for your ministries.

Reverend Sineta

Foreword

It is both an honor and a profound joy to offer this foreword for AI and the Black Church: Bridging Faith, Justice, and Innovation. This work emerges at a critical intersection in history where faith meets technology, and tradition encounters transformation. In these pages, Rev. Dr. Sineta Robertson courageously invites us into a conversation that many have hesitated to engage, yet one that we can no longer afford to ignore.

The Black Church has long stood as a beacon of hope, resilience, and prophetic witness in the face of adversity. From the days of slavery to the Civil Rights Movement and beyond, it has been a sacred space where faith fueled freedom, where wisdom guided justice, and where community nurtured survival and flourishing. Today, we find ourselves at another pivotal moment, one defined not by chains or segregation, but by algorithms, data, and artificial intelligence.

Dr. Robertson approaches this subject with both spiritual sensitivity and intellectual clarity. She does not present artificial intelligence as a threat to faith, but as a tool, one that, when guided by biblical principles and ethical discernment, can enhance ministry, expand outreach, and deepen our understanding of the world God has entrusted to us. At the same time, she does not shy away from the challenges. She names the ethical concerns, the potential for bias, and the risk of dehumanization, urging the Church to remain vigilant, prayerful, and rooted in truth.

What makes this book especially compelling is its call to action. It challenges leaders, pastors, and congregations not to retreat from innovation, but to engage it with wisdom. It reminds us that the same Spirit that empowered the early Church to navigate new cultural landscapes is present with us today, equipping us to navigate the digital age.

This work is not merely about technology, it is about stewardship. It is about how we, as the Body of Christ, responsibly engage the tools of our time to advance the Gospel, advocate for justice, and uplift our communities. Dr. Robertson's voice is both pastoral and prophetic, offering guidance that is as practical as it is inspiring.

As you read this book, I encourage you to do so with an open mind and a prayerful heart. Allow it to stretch your thinking, challenge your assumptions, and inspire your imagination. The future of the Church will not be shaped by those who resist change, but by those who discern how to faithfully navigate it.

Dr. Robertson has given us a timely and necessary resource. May it serve as a catalyst for thoughtful dialogue, faithful innovation, and Spirit-led transformation within the Black Church and beyond.

Grace and Peace,

Reverend Dr. Jamison Hunter

Senior Pastor, Mount Jezreel Baptist Church

Silver Spring, Maryland

Table of Contents

Introduction

The purpose of this book is twofold. First, it aims to give you a practical, accessible understanding of what artificial intelligence actually is and how it already shows up in your everyday life. Second, and more importantly, it invites you to explore the deeper question of how your faith should shape the way you engage with these technologies.

The Black Church has never faced a new challenge without first asking, *"What does God require of us here?"* This book is an attempt to apply that same question to the digital age. We will look at AI through a spiritual lens---examining its promises and its dangers, its potential for good and its capacity for harm---and we will ground those conversations in biblical values like compassion, stewardship, justice, and human dignity. This includes considering how advancements in AI impact ethical decision-making, community relationships, and the pursuit of justice.

By the end of this book, my hope is that you will feel not only informed but genuinely empowered. Empowered to lead your congregation through change. Empowered to ask hard questions of the technologies you use. And empowered to use every available tool---including AI---in service of the mission God has already placed in your hands. Ultimately, the goal is to empower faith communities to confidently navigate technological change while remaining anchored in their spiritual heritage and mission.

Chapter 1

What Is AI, or Artificial Intelligence?

Let's start with the basics, because this question matters more than it might seem. Artificial Intelligence, or AI, is not one single thing. It is a broad collection of tools and systems designed to simulate the kind of thinking and decision-making that we usually associate with human beings. Over the past several decades, AI has moved from the realm of science fiction into our pockets, our cars, our hospitals, and our homes.

Here is what makes AI feel both exciting and a little unsettling: most of the time, we do not even notice it. When you ask Siri or Alexa for directions, that is AI. When Netflix suggests a show you end up loving, that is AI. When your email quietly moves a suspicious message into your spam folder before you even see it---that is AI too.

Your phone's Face ID works because an AI system mapped the unique contours of your face and remembers them every time you try to unlock your screen. Your bank's fraud alerts fire because an AI noticed that a transaction did not match your usual patterns and flagged it before you were aware anything was wrong. Your smartwatch nudging you to stand up or drink more water? AI again, quietly watching out for your health habits. Navigation apps such as Google Maps and Waze use AI to analyze traffic data in real time, suggesting alternate routes to save you time and avoid delays. Autocorrect on your devices learns your typing style and common errors, helping you send clearer messages. These everyday examples show how AI is seamlessly integrated into our routines, enhancing convenience, security, and personal well-being.

For faith communities, and particularly the Black Church, understanding AI is more than just keeping up with technology; it is about discerning how these advancements align or conflict with spiritual values, ethical responsibilities, and communal well-being. By exploring the history, capabilities, and impact of AI, we are better equipped to thoughtfully navigate its opportunities and challenges. As church members use GPS apps to avoid traffic on the way to worship or receive health reminders from their smartwatches, they experience firsthand how AI can improve daily life. However, it is important to remain mindful of the ethical and spiritual dimensions these innovations present.

AI is neither a soul, a conscience, a prophet, nor a substitute for prayer or spiritual guidance. AI is a tool, not a person in charge.

Expanded Reflection: What AI Is and Is Not

Before we go any further, let us be clear about something important: AI is a tool, not a person. It has no soul, no conscience, no access to divine wisdom, and no ability to pray. It cannot replace pastoral leadership, prophetic voice, or spiritual discernment.

For ministry leaders, it is important to clearly define what AI can and cannot do. AI is capable of sorting information, making predictions, offering recommendations, and summarizing content by analyzing patterns in data. It can help with drafting text, organizing materials, and identifying key themes in large amounts of information. However, AI does not have spiritual authority, moral wisdom, covenant faithfulness, or the responsibility required of pastoral leadership.

A useful way to think about it is this: *AI acts like a very sophisticated calculator for language and patterns.* It can assist human work remarkably well. But it should never be allowed to replace the spiritual judgment, relational wisdom, and communal accountability that define faithful ministry.

As you consider how or whether to use AI in your context, a few practical questions are worth keeping close:

- ☐ What problem are we actually trying to solve?
- ☐ Which human relationships and expressions of care must remain central and irreplaceable, no matter what we adopt?
- ☐ What data does this tool use, and could that data harm someone if misused?
- ☐ How will we review the output for bias or errors before it reaches our congregation?

Chapter 2

AI You Use Every Day

Before we get into the theology and the ethics, it helps to look at what AI is already doing in ordinary daily life. Because here is the thing---AI did not arrive with a big announcement. It crept into our routines gradually, quietly, one helpful little feature at a time. By now, most of us are interacting with AI dozens of times a day without giving it a second thought.

Voice Assistants: Take voice assistants like Siri and Alexa. These systems use advanced speech recognition and natural language processing to understand not just what you are saying, but what you are likely trying to accomplish. They learn your preferences over time and adapt accordingly, which is why your assistant might already know that when you say "play my Sunday music," you mean something very specific. Over time, they learn your preferences and adapt their responses, providing more personalized assistance and streamlining daily tasks.

Home Security Systems: Modern home security systems are equipped with AI-powered features like motion detection, facial recognition, and automated alerts. These systems can distinguish between familiar faces and potential intruders, automatically notify you of unusual activity, and even integrate with smart locks or lighting to enhance safety and convenience. By combining voice assistants and intelligent security systems, many households experience a new level of connected living, where technology anticipates needs and safeguards well-being with minimal effort from the user.

Face ID: When you hold your phone up to unlock it, AI is running a real-time comparison between your current facial scan and a digital map it created of your face the first time you set it up. It can adjust for changes in your appearance---a new haircut, a new pair of glasses, a beard you grew over the winter---and still recognize you. The whole process takes less than a second. The underlying AI employs advanced algorithms to analyze key points on your face---such as the distance between your eyes, the shape of your cheekbones, and the contours of your jaw.

In addition to convenience, Face ID technology enhances security by making it extremely difficult for someone else to unlock your device using a photo or mask, as the system checks for depth and movement to ensure a real, live face is present. Many smartphones process this facial recognition data directly on the device rather than sending it to external servers, which helps protect your privacy. Over time, the AI continues to learn and improve its accuracy, adapting to gradual changes in your appearance while minimizing the chance of false positives. This blend of security, privacy, and ease of use demonstrates how AI is quietly woven into everyday routines, making technology both smarter and safer.

Autocorrect: Autocorrect is another quiet AI presence. It is not just checking your spelling against a dictionary---it is learning how you write, the words you use frequently, even the names of people and places you mention regularly. That is why your autocorrect eventually stops trying to "correct" the names of your church members.

When you type a message, autocorrect scans each word and compares it against a vast dictionary and statistical data on common spelling errors. For example, if you accidentally type "teh" instead of "the," autocorrect instantly recognizes the likely mistake and corrects it. But it does more than just fix typos---it can also anticipate entire words or phrases you commonly use, speeding up your communication with suggestions that appear as you type. Advanced autocorrect systems pay attention to the context of your sentences, so they are more likely to suggest "their" instead of "there" if it fits the overall meaning.

Over time, these systems adapt to your individual style, learning the slang, names, or technical terms you use frequently. If you repeatedly correct an autocorrect suggestion, the AI remembers your preference and becomes less likely to make the same mistake again. Autocorrect is especially useful for people who type quickly on small smartphone keyboards or who may have difficulties with spelling, ensuring their intended message comes across clearly. This everyday use of AI illustrates how technology can quietly enhance our daily lives, making digital communication smoother and more accessible for everyone.

GPS and Traffic Rerouting: Navigation apps like Google Maps and Waze are powered by AI that is constantly ingesting real-time data from thousands of other drivers, road sensors, and historical traffic patterns. These apps use GPS to keep track of where you are in real time and analyze data from other cars, road sensors, and past traffic trends. AI systems swiftly assess information about traffic jams, accidents, or road closures and recommend other routes to avoid delays. When it reroutes you three miles before a slowdown you did not even know was coming, that is AI at work.

Email and Spam Filters: Spam filters employ AI to automatically sort through the communications you receive. These filters check the subject lines, content, and sender information of each email. AI helps keep your inbox clean and organized by moving suspect emails to your spam or junk folder, looking for patterns in undesirable messages---such as suspicious links, particular keywords, or known spam addresses. These filters get better at spotting spam and hazardous content over time by watching how you deal with emails. They also keep you secure from scams and phishing attempts.

Bank Fraud Alerts: Your bank sends fraud alerts to let you know about behavior on your account that looks suspicious or might not be permitted. AI typically powers these notifications, looking for strange trends in your activities---such as large withdrawals, purchases in unfamiliar locations, or unexpected online activity. If the system sees something strange, it will send you a text, email, or app notification to make sure the transaction was really yours. This helps keep your money safe by letting you rapidly recognize and stop fraud.

Health Reminders: AI-powered health reminders are built into many smartphones and wearable devices. If you have been sitting too long, these reminders can tell you to get up, drink water, take your medicine, or walk. AI analyzes your everyday activities and trends to provide useful advice or timely nudges that help you develop better habits. For instance, your smartwatch might remind you to get up and exercise if you have been sitting for a long time, or it might send you a gentle reminder to drink enough water throughout the day.

Fitness Tracking: AI monitors activities such as walking, jogging, biking, and even sleep patterns. These devices keep track of your steps, heart rate, calories burned, and other health data. AI uses this information to provide you unique insights, help you create realistic objectives, and keep you motivated to continue being active. The system learns from how you work out over time and can give you personalized training tips, progress reports, and motivation to help you reach your fitness objectives.

AI didn't appear suddenly; it gradually integrated itself into our daily lives.

Expanded Reflection: Living with Everyday AI

Recognizing how thoroughly AI is already embedded in our routines should prompt us to slow down and ask some honest questions. As we reflect on the many ways AI is woven into our daily lives, it becomes clear that these technologies are more than just convenient tools---they are shaping how we work, connect, and care for ourselves and one another. From the moment our alarms go off to the instant we receive an alert about a potential security concern, AI is quietly present, helping us navigate an ever-changing world with greater ease and safety.

This seamless integration, however, invites us to pause and consider the deeper implications. Each time we use AI for directions, filtering emails, or health reminders, we are entrusting a piece of our lives to systems that others have built, trained on vast amounts of data, and that are constantly evolving. While these tools often serve us well, it is important to remember that they are not neutral or infallible. They reflect the values, priorities, and limitations of those who design and program them.

For faith communities---particularly the Black Church, which has always had reason to think carefully about who controls information and how it is used---the rapid normalization of AI warrants more than passive acceptance. How do these technologies impact our relationships, our sense of identity, and our communal responsibilities? Are they helping us live with greater dignity and wisdom, or do they sometimes distract or divide us? As we become more accustomed to digital assistance, how do we ensure that the core values of love, justice, integrity, and care for others remain central in our lives?

The everyday presence of AI invites us not only to appreciate its benefits but also to remain vigilant and discerning. We should feel empowered to ask hard questions about privacy, fairness, security, and the potential for bias. We should advocate for technology that uplifts and protects all people, especially those who are most vulnerable or excluded. And we should remember that, while AI can enhance our routines, it cannot replace the wisdom, compassion, and spiritual insight that come from genuine human communities.

In the chapters ahead, we will continue to explore how faith can guide our engagement with technology, encouraging us to use AI thoughtfully, responsibly, and with hope. Let us move forward not only as users of these tools but also as active stewards, shaping a future where technology serves our highest values and the common good.

As we reflect on the many ways artificial intelligence is woven into our daily lives, it becomes clear that these technologies are not distant concepts but constant companions. From navigation systems to financial alerts, from sermon research tools to communication platforms, AI is already shaping how we move, think, and serve.

The question is no longer whether the Church will encounter artificial intelligence. The question is how we will interpret it. Will we approach it with fear? Will we embrace it uncritically? Or will we engage it with theological depth and spiritual discernment? The Black Church has never navigated new systems without first seeking a biblical framework. Before determining how to use artificial intelligence, we must understand how God operates within flawed human systems. Scripture offers us such a framework---not from the modern digital age, but from an ancient imperial context.

Before we build policies for AI, we must build theology for AI.

Chapter 3

Genesis 50:20 -A Guide for Making Decisions

There is a verse of Scripture that I keep coming back to as I think about artificial intelligence and the role the Church is called to play in this moment. It comes near the end of the Joseph narrative in Genesis, in the conversation between Joseph and the brothers who had sold him into slavery years before. His response to them is one of the most theologically profound statements in all of Scripture:

"You meant evil against me, but God meant it for good, to bring it about that many people should be kept alive, as they are today."
(Genesis 50:20, ESV)

Joseph's declaration is one of the clearest biblical affirmations of divine sovereignty operating alongside human agency. In today's world, where emerging technologies like artificial intelligence can serve both beneficial and detrimental purposes, this biblical perspective holds particular relevance. Genesis 50:20 reminds faith communities and the Black Church to approach new tools and innovations with both discernment and hope. We are called to look beyond the surface, recognizing that while people may misuse technology for selfish or destructive ends, God can still work through these circumstances to accomplish something positive for the community and the world.

What Joseph is describing here is something theologians call **concurrent agency**, the reality that two completely different wills can be operating within the same event at the same time. His brothers acted out of jealousy and malice. God acted out of providence and purpose. Same event, same chain of history. Two entirely different intentions at work.

This is not dualism. God was not simply "reacting" to what the brothers did, improvising a response after the fact. Scripture says God "meant" it for good. That word is intentional. This presents a profound theological reality: ***Two Intentions One Event.*** Human beings acted freely and sinfully. God sovereignly worked through their actions for redemptive good. This principle is divine providence. God is not absent within history. For the Black Church, this is not abstract doctrine. It is lived experience.

Applying this scripture means asking thoughtful questions: Who is benefiting from this technology? Who might be harmed? Are our choices motivated by love, justice, and care for others? By reflecting on the intentions behind both the creation and use of AI, faith leaders and congregations can better determine how to use these tools in ways that align with their spiritual values. This approach does not ignore the risks or challenges such as privacy concerns, bias, or the potential for exclusion but rather acknowledges them and seeks to respond in ways that honor God's purpose.

Naming Evil Without Surrendering Hope

One of the most powerful gifts this passage offers is permission to tell the truth. Genesis 50:20 does not ask us to pretend that harm is not harm or that evil is not evil. Joseph did not minimize what his brothers had done. He named it clearly: *"You meant evil against me."* Only after that honest acknowledgment did he speak about God's redemptive purpose.

The Black Church has always understood this tension. Enslavement was meant for exploitation. Segregation was meant for control. Surveillance was meant for suppression. And yet, through those very systems, God preserved faith, nurtured resilience, and raised up prophetic voices that changed the arc of history. Genesis 50:20 does not teach us that evil is acceptable. It teaches us that evil does not get the final word.

We must bring that same clarity to artificial intelligence. AI systems can reinforce racial bias. They can enable invasive surveillance. They can deepen economic inequality. They can substitute efficiency for relationship in ways that leave people feeling more isolated, not more connected. These are real dangers, and they must be named. But naming the danger does not require abandoning hope. The same technological systems that harm can also expand access to education, strengthen ministry communication, increase accessibility for people with disabilities, support pastoral administration, and help preserve the digital records of our history and testimony.

The theological question becomes not "Is technology good or evil?" but "How will it be used?"

Sovereignty Does Not Remove Responsibility

Here is what Genesis 50:20 does not say: it does not say that because God can bring good from a broken situation, we are off the hook for our choices within that situation. Joseph's brothers were still accountable for what they had done. The fact that God redeemed the outcome did not excuse the wrongdoing. Redemption is not endorsement.

The same principle applies to us. We cannot throw up our hands and say, "God will work it out," and use that as a reason to disengage from the hard ethical work of evaluating the technologies we use. Scripture never teaches passive faith. It calls us to active righteousness. God's sovereignty invites courage, not complacency. When it comes to AI, that means asking difficult questions:

- Who designed this system, and what values were baked in from the start?
- Whose data was used to train it, and were those people given a voice in how it was used?
- Who benefits economically from this technology?
- Who might be excluded or harmed by the way it operates?
- What safeguards are in place, and who is responsible for maintaining them?

Redemption Is Not Endorsement

The fact that God can bring good from a system does not mean the system is inherently just. Joseph's elevation did not sanctify every Egyptian policy. It allowed him to mitigate harm within it. Similarly, AI's usefulness does not make it morally neutral. Technology reflects the values of those who build it. If unexamined, it can reinforce injustice. If stewarded wisely, it can expand care and efficiency without replacing shepherding.

The Church must resist two extremes: technological fear and technological triumphalism. Genesis 50:20 offers a third way: faithful engagement.

The Moral Test: Does It Preserve Life?

Joseph himself gives us a practical test for evaluating whether a system serves God's purposes. He says the point of everything that had happened was *"to save much people alive."* Preservation of life. Protection of the vulnerable. Restoration of community.

That is our test for AI as well. When we are evaluating a new technology or deciding whether to use an existing one, we can ask: Does this preserve human dignity? Does it strengthen relationships rather than replace them? Does it protect privacy? Does it promote justice? Does it serve those who are most vulnerable in our community? If the answer to those questions is yes, then it may be used responsibly. If the answer is no, it must be reshaped or rejected.

A Commissioning Verse

Genesis 50:20 is not merely comfort. It is commissioning. Joseph did not simply survive betrayal; he stepped into leadership to prevent future harm. Likewise, the Church is not called merely to endure artificial intelligence; it is called to lead within it. Lead with wisdom, ethical clarity, and unwavering moral vision. The God who worked through empires can work through algorithms. But often, God works through faithful people willing to use systems responsibly.

The problem isn't the tool itself, but who is using it.

Expanded Reflection: Using Genesis 50:20 for Decision-Making

Genesis 50:20 encourages us to be realistic about the possibility of harm while trusting in God's power to bring good out of difficult situations. When deciding how to use AI, the Black Church can approach the process by saying, "We acknowledge the risks, but we also hold onto hope." This means evaluating the intentions behind the technology, considering its impact, and ensuring accountability.

Here is a practical decision-making framework that ministry teams can use when evaluating any new AI tool:

1. **Begin with prayer.** Bring the decision before God before you bring it before a committee.
2. **Name the intended benefit clearly.** What problem is this tool designed to solve?
3. **Identify who might be at risk.** Who could be harmed, excluded, or disadvantaged by how this tool operates?
4. **Put safeguards in place.** Do not just adopt the tool, establish who is responsible for monitoring it.
5. **Make decisions with accountability.** No one person should carry this alone.
6. **Review the results after 30 to 90 days.** Does the outcome align with your values and your mission?

Chapter 4

Black Church Historical and Theological Witness

The theological truths of Genesis 50:20 are not abstract ideas. They have been lived out in the history of the Black Church for centuries. The declaration that "You meant evil against me, but God meant it for good" has functioned not simply as comfort in hard times, but as an interpretive lens, a survival theology, and a source of prophetic fuel across generations.

Before we examine the ethical dangers and opportunities of artificial intelligence, we need to understand the tradition we are bringing to this conversation. The story of the Black Church is the story of living inside structures that were not designed for our flourishing and finding ways to transform them through resilient faith. That history is not just inspiration. It is practical wisdom for this moment. The Black Church has always lived in the tension of Genesis 50:20.

Faith Under Surveillance: The Invisible Institution

During slavery, faith was not freely practiced. It was monitored, restricted, and sometimes deliberately distorted by enslavers who understood that a person's spiritual life shapes their sense of identity and possibility. And yet, enslaved Africans found ways to carve out sacred space beyond the reach of surveillance, in what historians have called the "invisible institution" hush harbors hidden in woods, brush arbors, and secluded clearings far from the master's house.

These were not merely secret meetings. They were theological acts of resistance. In those spaces, enslaved believers interpreted Scripture through the lens of deliverance. The Exodus narrative became deeply personal. The prophets became urgently relevant. Songs encoded hope in melody: "Steal Away," "Go Down, Moses," and "Swing Low, Sweet Chariot" were not simply hymns, they were declarations that God saw, God heard, and God would act.

Surveillance was real. Oppression was real. The system was designed for control. And yet God sustained a people in the middle of all of it. Genesis 50:20 was being lived, not just preached, in the soil of those gatherings. Evil was meant for exploitation. But God preserved a people.

The Sanctuary as Strategy: Faith During Segregation

After slavery, new systems of control emerged. Jim Crow laws institutionalized racial hierarchy. Economic barriers limited advancement. Political exclusion silenced voices. But the Black Church did not retreat. It expanded its role and deepened its commitment to community life.

Sanctuaries became classrooms when education was denied. Church basements became planning centers for civil rights action. Pulpits became platforms for prophetic proclamation. The church was not merely a refuge from injustice; it was a command center against it. Leaders like Dr.

Martin Luther King, Jr., and countless unnamed pastors and lay leaders grounded their activism in theology. Their vision of justice was not political invention but biblical conviction. Scripture shaped strategy. Prayer preceded protest. Worship sustained courage.

The 16th Street Baptist Church in Birmingham, Alabama, stands as a painful and powerful reminder of both the cost and the courage of faith-driven resistance. Even when violence targeted sacred space, the Church did not abandon its witness. Again, Genesis 50:20 was embodied: systems intended exclusion. God cultivated liberation.

Navigating Complicity and Calling

The historical witness of the Black Church is not without complexity. At times, survival required compromise. At times, institutional preservation created tension with prophetic confrontation. At times, churches navigated systems carefully to avoid destruction. These tensions are not a source of shame they are evidence of wisdom under pressure.

Even in those moments of limitation, the Black Church consistently returned to its core calling: "To seek justice, love mercy, and walk humbly with God." This pattern reveals something critical for our present moment: engaging with systems does not automatically equal endorsing them. Joseph served in Egypt without sanctifying Egypt. The Black Church functioned within America without endorsing its injustices. Likewise, engaging digital systems such as artificial intelligence does not mean surrendering theological integrity. Discernment has always been the key.

A Theology of Survival and Stewardship

The Black Church did not merely develop a theology of survival. It developed a theology of stewardship within adversity. From mutual aid societies and burial funds to scholarship programs and health clinics, congregations stepped into every gap left by unjust systems. When public institutions failed, churches organized. When economic access was restricted, churches pooled resources. When political voice was silenced, churches amplified it.

This was not accidental activism. It was theological response. If God could bring good from betrayal, then the Black Church could cultivate hope inside hardship. If divine sovereignty was trustworthy, then communal responsibility was necessary. Genesis 50:20 did not produce passivity, it produced participation.

Annual homecomings and Jubilee celebrations remind us that memory matters. They connect past perseverance with present responsibility. They testify that the God who sustained our ancestors remains active today. The same theological courage that sustained the Black Church through slavery and segregation must now guide it through automation and artificial intelligence.

Lessons for the Digital Age

Today, the Black Church faces a new kind of systems not plantation economy or Jim Crow law, but digital infrastructure. Artificial intelligence, algorithmic decision-making, data surveillance, and automation now shape daily life in ways that are often invisible and deeply consequential. The parallels to our history are instructive and sobering.

Earlier generations navigated physical surveillance with spiritual clarity and communal solidarity. The task before us now is to do the same with digital surveillance. This does not mean withdrawing from technology, any more than it meant withdrawing from the political systems of the Civil Rights era. It means engaging actively, guided by theological conviction and moral clarity naming harm honestly, building structures of accountability, and ensuring that technology serves the people rather than the other way around.

Technology, Power, and Prophetic Responsibility

The Black Church understands power, how it operates, how it excludes, and how it can be challenged. This historical awareness uniquely equips us to approach artificial intelligence both critically and constructively. Technology is not neutral. It reflects the priorities, assumptions, and economic interests of those who design it. Without scrutiny, it can reinforce inequality. Without oversight, it can marginalize the vulnerable.

But history also teaches us that systems can be shaped. Just as the Black Church used print media, radio, and television to expand the gospel witness in earlier generations, digital platforms now extend ministry reach. Worship services stream across geographic boundaries. Bible studies convene online. Social justice campaigns mobilize through digital networks. Technology can amplify mission when it is guided by moral clarity.

The Enduring Witness

The Black Church has endured slavery without surrendering identity. It has confronted segregation without abandoning faith. It has navigated political systems without forfeiting prophetic voice. This history offers more than inspiration; it offers practical guidance for how to engage powerful systems without losing yourself in them.

Genesis 50:20 continues to echo across generations: evil may intend harm, but God intends preservation. The Church stands in that tension, not denying injustice, not surrendering hope, but stewarding faithfully within every era. The systems may change. The sovereignty of God does not. And the calling of the Church remains the same: to share the gospel and to bear witness to justice, dignity, mercy, and hope in every age.

Expanded Reflection: Historical Wisdom for Modern Systems

The history of the Black Church teaches us how to stay true to our faith while engaging and transforming unfair systems. For example, hush harbors and secret worship gatherings during slavery highlight the importance of privacy and spiritual independence. The Civil Rights movement demonstrates that technology and organization can enhance freedom, but only when they align with a robust moral vision.

This legacy guides our approach to digital ethics today: we honor the personal stories of congregants as sacred, protect vulnerable people from unwanted exposure, challenge systems that exploit others for profit, and use technology to expand care, learning, and justice.

One practical step for church leaders is to include a short "history moment" in conversations about technology, linking current concerns about data and privacy to past experiences with surveillance. This helps congregations see why privacy and consent remain vital topics today.

Chapter 5

Actual Dangers of AI

It would be dishonest to write a book about AI and the Church without spending real time on the dangers. Artificial intelligence's increasing integration into daily life raises pressing questions and real concerns about its impact on faith communities, particularly the Black Church. While technology presents exciting opportunities for ministry and outreach, it also presents a range of potential dangers that require careful consideration. From threats to privacy and the risk of reinforcing biases to the subtle erosion of authentic relationships and reliance on technology over spiritual discernment, these challenges demand careful pastoral reflection and leadership.

Artificial intelligence is not benign. It is not automatically a force for good just because it is useful. And the Black Church, with its long memory of navigating systems that promised convenience or progress while quietly reinforcing harm, is in a particularly good position to look at this technology with clear eyes.

Loss of Privacy

Privacy is not just a preference. For communities that have historically been targeted by surveillance monitored, catalogued, and controlled, privacy is a matter of safety and dignity. Loss of privacy means that people might not know or agree to having their personal information, actions, and communications gathered, watched, shared, or made public. Many congregants do not approve of having their images posted on social media platforms and the like.

AI systems typically require vast amounts of data to function: information about who we are, how we behave, what we believe, where we go, and who we talk to. When that data is not collected transparently, stored securely, or used only for the purposes we agreed to, we have a privacy violation on our hands.

For the Black Church, this concern is especially acute. Many congregants share deeply personal information in the context of ministry, health struggles, family conflicts, financial hardship, spiritual crises. If AI-powered systems used for church administration, outreach, or online ministry gather more information than is necessary, or if that information is not maintained securely, members' privacy and trust may be violated. There have been documented instances of surveilling or weaponizing the data of nonconformists and minority communities, making this matter especially concerning for faith communities with a history of being monitored. All organizations, even religious groups, need to be careful about how AI technologies use data ensuring transparency, obtaining permission, and following strict security rules to protect everyone's privacy and dignity.

Prejudice and Exclusion

Bias and exclusion describe how AI systems can unintentionally make unfair treatment of some groups worse or more common. AI systems learn from data that comes from the actual world. If that data contains stereotypes, biases, or reflects current societal disparities, the AI may learn and reproduce those patterns. This is not theoretical---it has already shown up in facial recognition technology that performs poorly on darker skin tones, in hiring algorithms that screen out qualified candidates from certain zip codes, and in health care AI that produces less accurate diagnoses for Black patients.

For instance, if an AI tool is used for church outreach or resource allocation but is mostly trained on data from one demographic group, it can miss or distort the needs and voices of other groups, leaving out populations that are already on the margins. For the Black Church and communities that have historically been underrepresented or misrepresented in institutional data, this concern deserves serious attention. If AI-driven decision-making is not watched closely, it might deepen existing inequalities by favoring some voices over others. To address these concerns, it is important to use diverse, representative data when building AI systems, to check results for fairness on a regular basis, and to ensure that human judgment and responsibility lead how technology is used.

Efficiency Taking the Place of Relationship

Here is a subtle danger that may be the one that most easily slips past our defenses: AI can make the church more efficient in ways that slowly erode the relational culture that makes the church the church. AI solutions can make things like church administration, communication, and even pastoral care easier and faster. But if you rely too much on the technology, you may diminish the personal interaction that makes ministry meaningful.

Automated follow-up messages can go out on time, but they cannot replace the phone call from a deacon who noticed you were not in service last Sunday. AI-driven counseling may lack the empathy, comprehension, and cultural sensitivity inherent in genuine interpersonal relationships. In the Black Church, relationships have always been the basis for resilience, activism, and spiritual growth. When efficiency becomes the primary goal, members can begin to feel like data points rather than people. The warmth, the accountability, the sense of being truly known and seen, these things cannot be generated by an algorithm. They have to be cultivated by people who are genuinely invested in one another.

To prevent efficiency from overriding relationship, church communities need to employ AI carefully. Technology can help, but it should never take the place of the important human connections that keep spiritual and community growth going.

Technology Taking the Place of Discernment

Perhaps the deepest danger is this: that we might gradually stop doing the hard spiritual work of discernment and substitute algorithmic recommendations in its place. There is a rising risk that the use of technology, including artificial intelligence, may slowly take the place of the important practice of spiritual and moral discernment as these tools become more common in the church.

Discernment means prayerfully seeking God's will, weighing the counsel of wise people, listening to the Spirit within community and it is not a process that can be

outsourced. It is irreducibly human, irreducibly spiritual, and irreducibly communal. When technology is used to automate or simplify difficult decisions like choosing ministry priorities, allocating resources, or even interpreting Scripture. There is a risk that the church will start to rely on algorithms and automated suggestions instead of doing the spiritual and communal work of seeking God's guidance.

In the Black Church, discernment has historically been essential for overcoming challenges, promoting justice, and cultivating resilience. For instance, an AI tool might recommend ways to reach out to people based on data trends, but it might not take into account the specific requirements or spiritual dynamics of a particular group. Such an outcome might result in decisions that are effective yet devoid of depth, empathy, and alignment with core beliefs.

To keep technology from displacing discernment, it is important for faith leaders and the church to perceive AI as a tool that *assists* in seeking God's will, not as a replacement for it. This means putting prayer, thoughtful conversation, and communal reflection ahead of technological fixes. By maintaining this balance, technology has the potential to strengthen faith-based discernment rather than diminish it.

Expanded Reflection: Risks That Require Pastoral Leadership

Each of these dangers becomes more concrete when illustrated with examples specific to ministry settings:

1. Privacy concerns arise when a church database or online form collects more personal information than is truly necessary.

2. Bias is evident when automated tools or recommendations exclude certain neighborhoods, age groups, or language communities.
3. Relational erosion happens when automated replies replace genuine pastoral follow-up.
4. Discernment loss happens when leaders begin deferring to what the technology suggests rather than doing the slower, richer work of prayer and communal conversation.

A practical response to all of these risks is to adopt a "human-in-the-loop" approach: AI can be used to draft, organize, or suggest, but a designated leader must always review, edit, and approve everything before it reaches the congregation.

Chapter 6

The Black Church's Use of AI for Redemption

Having named the dangers honestly, we can now turn to the other side of the ledger and there is a real other side. Rather than viewing artificial intelligence as a threat, many congregations are beginning to explore its redemptive possibilities, harnessing AI tools to enrich worship, strengthen community bonds, and advance social justice. Artificial intelligence, used wisely and with clear theological boundaries, can genuinely serve the mission of the Black Church.

Help with Sermon Research

One of the most time-consuming parts of pastoral ministry is research. Preparing a theologically rich, culturally relevant, and scripturally grounded sermon requires sifting through commentaries, historical resources, contemporary context, and the specific needs of your congregation. Ministers and church leaders can quickly search through biblical texts, theological commentaries, and historical resources by using AI-powered search, language analysis, and content summarizing tools. This assistance can help pastors discover fresh ideas, locate pertinent scriptures, and connect with a diverse array of viewpoints including those from various traditions and backgrounds that may not have been previously accessible.

In the Black Church, AI can help find stories of resilience, justice, and advocacy from both modern and historical sources. This can make sermons more captivating by providing examples that are relevant to the culture. By utilizing advanced search capabilities, language models, and data analysis, AI tools can sift through vast archives of oral histories, written testimonies, and digital media to uncover narratives that might otherwise remain hidden, powerful personal accounts from the Civil Rights Movement, spotlights on contemporary leaders in social justice, or examples of local community efforts that embody the values of resilience and advocacy.

Additionally, AI can help tailor these stories to the specific context of each congregation, ensuring that sermon illustrations resonate with the lived experiences of church members. This not only enriches preaching but also strengthens communal identity and inspires congregations to continue their legacy of justice and hope.

Churches need to think carefully about fairness and representation when using these tools. You should regularly check AI-generated content for bias, seek feedback from people of different backgrounds, and make cultural understanding a priority. AI-generated content should always be reviewed before use, checked for theological accuracy, and assessed for cultural sensitivity. The AI does the research. The pastor does the discernment. That distinction must never blur.

Help with Bible Study Preparation

AI tools can also support Bible study preparation by helping leaders quickly compare translations, locate commentary, generate discussion questions, and identify thematic connections across Scripture. Faith leaders and participants can rapidly find key Bible verses, compare multiple translations, and read a wide range of commentaries and scholarly views by using AI-powered tools. These tools can help you organize your study materials, come up with discussion questions, and find themes or connections between biblical texts that you might not have seen otherwise.

For Black Church communities and other culturally rich spiritual traditions, AI can help find information and perspectives that match their histories, experiences, and theological focuses, making sure that Bible study sessions are not only instructive but also relevant to the lives and spiritual journeys of the people who attend. AI may also make study more inclusive and participatory by suggesting resources from various viewpoints and traditions, which can lead to richer knowledge and conversation.

For smaller congregations or volunteer leaders who may not have formal theological training, this kind of assistance can make quality Bible study preparation more accessible. The goal is to use AI to support the preparation process not to replace the communal interpretation, prayerful reading, and shared wisdom that make Bible study genuinely transformative. Technology should make the preparation easier so that the conversation can go deeper.

Help with Administration

Church administration is often invisible and frequently exhausting. By automating regular and time-consuming tasks,

AI can make things much easier for churches and administrative staff. These tasks could involve keeping track of attendance, making and sending out newsletters, monitoring membership databases, arranging meetings, and handling financial reports. By making these tasks easier, AI gives church leaders, staff, and volunteers more time to focus on mission-driven activities like spiritual mentorship, pastoral care, and community outreach.

Administrative work in the Black Church can be especially challenging due to limited resources and the wide range of ministry programs that must be managed. AI-powered tools offer practical support by helping leaders plan events, communicate effectively with congregants, and streamline daily operations. These technologies can simplify recordkeeping, generate announcements for upcoming events, reduce paperwork, and minimize errors. Automated scheduling tools help coordinate ministry teams more efficiently, while AI-driven analytics provide insights into community needs allowing leaders to make more informed decisions and steward resources more effectively.

There are many benefits to using technology to help with administrative tasks, but it is vital to use these tools carefully. Churches should ensure that automation enhances rather than replaces members' sense of connection and personalization. Rather than replacing human connection, AI serves as a supportive tool, helping ministries operate more efficiently, stay better organized, and remain focused on people and purpose.

Help with Outreach and Follow-Up

AI can significantly enhance outreach and follow-up efforts within the Black Church. By leveraging AI-powered tools, churches can more effectively identify and connect with new visitors, engage members who may have become less active, and ensure that no one feels overlooked. For instance, AI can analyze attendance patterns and communication records to automatically generate personalized follow-up messages, reminders, or invitations to upcoming events. These automated systems can help church staff prioritize outreach to those who may need additional support or encouragement.

Additionally, AI-driven platforms can segment congregational data to tailor outreach strategies for different groups such as youth, seniors, or new members making communications more relevant and impactful. AI chatbots and virtual assistants can provide real-time responses to inquiries from community members, offering information about church services, resources, or volunteering opportunities. AI can also assist in making messages more relevant to the community by highlighting important ministries and offering resources that address social justice concerns. Follow-up tools can make it easier to gather feedback, set up small groups, and coordinate care for members who are having a difficult time.

AI can make outreach programs far more effective and far-reaching, but technology should never take the place of real human interaction. Faith leaders can use AI as a tool to support and strengthen their relationships with people, not replace them. The key is always the same: AI surfaces the information, and human beings do the caring. By carefully using AI, churches can create a sense of belonging and care for everyone in their community.

Inclusion and Accessibility

Some of the most compelling applications of AI in ministry are in the area of accessibility. AI creates new opportunities for churches to intentionally design ministries that are accessible to all people, regardless of physical ability, language, age, or educational background. Rather than simply improving access to information, AI can help churches remove long-standing barriers that have often limited full participation in worship and discipleship.

For example, AI-enabled tools can provide real-time captioning during worship services, allowing individuals who are deaf or hard of hearing to fully engage. Language translation technologies can support diverse congregations by offering sermons, Bible studies, and church communications in multiple languages simultaneously. AI can also support neurodiverse learners by adapting content into different formats, such as simplified text, visual summaries, or interactive teaching aids. Digital assistants and AI-powered platforms can extend ministry beyond the physical church building, so that homebound members, seniors, or those with health challenges can access sermons, devotionals, and prayer support in ways that are personalized and easy to navigate.

When a ministry uses AI to ensure that a deaf congregant can fully participate in worship, or that a Spanish-speaking family can access the same teaching resources as everyone else, it is not simply using technology cleverly. It is living out the gospel call to welcome everyone to the table. Ultimately, when used thoughtfully, AI becomes a tool for equity in ministry helping the church live out its calling to be inclusive, welcoming, and responsive to the needs of all people.

AI can help in ministry, but it should never take the place of guidance from the Holy Spirit.

AI is useful, but it cannot replace the church's vital role of shepherding. Real shepherding means having personal relationships, giving spiritual direction, being empathetic, and truly knowing what the people in the community have been through. Listening, understanding, and walking with people through good times, bad times, and times of growth are the most important parts of ministry. Being truly present with others gives you spiritual understanding, compassion, and depth in your relationships that technology cannot match.

So, faith leaders should utilize AI carefully, as a tool that helps their ministry, not as a replacement for the personal touch that is so important in shepherding. Pastors can ensure that their sermons and ministries stay biblically rooted in love, caring, and the guiding principles of their religious tradition by balancing technical support with spiritual reflection, cultural knowledge, and real connection.

AI can help in ministry, but it should never take the place of shepherding God's sheep.

Expanded Reflection: Redemption in Practice

The most meaningful way AI can be used redemptively in ministry is not just about adopting new technology; it is about increasing the church's capacity to care for people. When AI helps save time on administrative tasks, that extra time should be intentionally directed toward core shepherding work: visiting those in the hospital, having discipleship conversations, mentoring youth, providing comfort during grief, and being present in the community.

Set a clear boundary: AI can support research and planning but should never replace genuine pastoral care. True pastoral care demands accountability, confidentiality, and authentic relationships qualities that technology cannot provide.

It is a beneficial practice to write an **"AI Use Statement"** for your ministry. This should be a short paragraph describing AI's uses (like drafting newsletters or organizing notes), its non-uses (like counseling or personal pastoral responses), and how members' privacy is protected. This statement builds trust with your congregation and sets clear expectations for everyone on staff.

Chapter 7

Understanding AI Prompts - Speaking Wisely to the Tool

AI does not think on its own. It responds to instructions. A "prompt" is the question, command, or direction you give an AI system, and the quality of what the AI produces depends almost entirely on the quality of what you ask. Artificial intelligence does not possess independent thought, wisdom, or intention. Those inputs, known as *prompts*, are the questions, commands, or instructions given to an AI system that shape the responses it produces. In many ways, prompts serve as the bridge between human intention and technological output.

Understanding prompts, therefore, is not merely a technical skill, it is an act of stewardship. The words we choose, the context we provide, and the purpose we set all determine how effectively and responsibly AI can be used in ministry and daily life. In ministry settings, understanding prompts is essential. The quality of what AI produces depends heavily on the clarity, depth, and intention behind what is asked.

What Prompts Look Like

A prompt can take many forms:

- A question: "Summarize Genesis 50:20 and its theological implications."
- A request: "Draft a Bible study outline on forgiveness for adult learners."
- A command: "Create five discussion questions for a youth ministry session on identity."
- A refinement: "Rewrite this sermon introduction with a Black Church historical lens and a more pastoral tone."

AI does not interpret emotion, divine inspiration, or spiritual conviction. It interprets language patterns.

The more thoughtful and specific the prompt, the more useful the response. In other words, AI reflects what you feed it.

Why Prompts Matter in Ministry

For the Black Church, prompts are not merely technical instructions, hey are stewardship decisions. The way we ask shapes what we receive. Consider the difference between these two prompts:

A weak prompt:

"Write a sermon about hope."

A stronger prompt:

"Write a sermon outline on Genesis 50:20 emphasizing God's sovereignty, Black Church resilience, and practical applications for a congregation navigating economic hardship."

The second prompt gives the AI theological direction, cultural context, and a specific audience. The result will be incomparably more useful for actual ministry. Prompts matter because they determine theological framing, influence cultural representation, shape tone and sensitivity, reduce bias and misunderstanding, and protect pastoral integrity.

AI is only as responsible as the person guiding it.

Principles for Writing Wise Prompts

1. **Be Specific.** Vague instructions produce generic results. Specific prompts produce contextual results.
2. **Provide Context.** Include your audience (pastors, youth, lay leaders), cultural lens (Black Church tradition), and purpose (training, sermon prep, outreach).
3. **State Theological Boundaries.** If content must align with Scripture, say so. If it must reflect Baptist theology or a particular doctrinal tradition, state it clearly.
4. **Require Review.** *AI-generated content must always be reviewed before use. Writing a good prompt does not remove pastoral responsibility.*
5. **Protect Privacy.** Never enter confidential counseling information, private member data, or sensitive financial details into AI systems.

Ministry Prompt Examples

Below are examples of responsible ministry prompts:

Sermon Preparation:

"Create a three-point sermon outline on Micah 6:8 incorporating themes of justice, mercy, and Black Church historical advocacy."

Bible Study:

"Generate five open-ended discussion questions on Romans 12:1-2 appropriate for adult learners."

Administrative Help:

"Draft a church newsletter announcement about our upcoming health fair using warm and pastoral language."

Leadership Training:

"Summarize key ethical concerns of AI use in churches and provide discussion questions for ministry leaders."

These prompts guide the tool while maintaining pastoral oversight.

What Prompts Should Not Do

AI prompts should never:

- ☐ Ask AI to provide spiritual direction in place of prayer.
- ☐ Replace pastoral counseling.
- ☐ Interpret personal prophecy or provide a word from the Lord.
- ☐ Make final theological determinations.
- ☐ Generate confidential member communications without human review.

AI can assist with preparation. It cannot shepherd souls.

Prompting and Discernment

Prompting can be understood as a modern form of stewardship. Just as pastors prepare prayerfully and thoughtfully before delivering a sermon, leaders must also be intentional in how they engage AI. In this sense, prompting echoes the wisdom of James 1:5: *"If any of you lacks wisdom, let him ask of God..."*

Before turning to AI for assistance, we first seek God for wisdom. Prayer should shape our questions, Scripture should guide our thinking, and spiritual discernment must be used to evaluate every response.

A Practical Framework for Churches

When using AI prompts in ministry:

1. Pray before prompting.
2. State the ministry purpose clearly.
3. Include theological and cultural context.
4. Review and edit thoroughly.
5. Take full responsibility for what is shared.

AI does not carry spiritual accountability. Leaders do.

Chapter 8

Ethical Guidelines

As Black churches and faith communities use AI in their work to spread the gospel and help others, it is important to set clear ethical boundaries so that technology supports the spiritual goal instead of getting in the way. These guidelines protect the church's ideals and the health of its people by providing recommendations for how to utilize AI in a responsible, respectful, and spiritually sensitive way.

While artificial intelligence provides invaluable tools for enhancing outreach, inclusion, and operational efficiency within faith communities, it is necessary to acknowledge the irreplaceable value of human leadership and spiritual discernment. AI can streamline processes and provide access to resources, yet true pastoral care, empathy, and guidance must always come from church leaders guided by faith and the Holy Spirit.

Human Authority Always Overrides AI

AI is never more powerful than people.

This is the foundational principle: AI is a tool, not a leader. Every AI-generated recommendation, draft, analysis, or suggestion must be reviewed, evaluated, and ultimately approved or overridden by a human being who is accountable to the congregation and responsible to God.

This matters especially for decisions that affect people's spiritual lives, relationships, or well-being. An AI might flag someone as a "low engagement" member based on attendance data without knowing that they have been sitting with a dying parent for three months. A pastor who knows that family will respond very differently than an algorithm that does not. Those who are accountable to their spiritual foundation must always make decisions that affect the church community, its teachings, or its members.

In all areas of ministry, it is essential that human leadership remain superior to artificial intelligence. Ministers, pastors, and church elders must always be the ones who interpret Scripture, make pastoral decisions, and give spiritual advice. In practical terms, AI-generated material, suggestions, or analyses should be viewed as aids and sources of information, not as definitive answers or directives. Faith leaders must examine, evaluate, and, where warranted, override AI findings particularly when such outputs may conflict with the church's fundamental beliefs, theological principles, or the actual experiences of its members.

Prayer Comes Before Production

Before you make something using AI, pray.

Before starting, installing, or executing any AI-driven projects in church ministry, it is important to pray and seek God's guidance first. Prayer is a way to bring the mission of the church back to God and focus on His guidance, looking for heavenly wisdom instead of technological efficiency or novelty. When church leaders and members pray first, they ask the Holy

Spirit to help them plan, decide, and ensure that their technical efforts align with God's Word and the church's spiritual goals.

Praying before you prompt means approaching every use of AI with intentionality and humility. It means acknowledging that technology does not have all the answers, that spiritual discernment cannot be automated, and that the mission of the Church is ultimately God's mission. This approach acknowledges the potential of AI to offer valuable tools and support yet affirms that the true foundation of ministry lies in spiritual connection, empathy, and discernment.

"Prayer comes before production" is a reminder for churches to keep their spiritual concerns in mind. It says that before using new technologies, we should take thoughtful pauses, open our hearts, and work together to figure out what to do. It also says that every ministry, even one that incorporates AI, must be based on faith, trust, and a dedication to serving God and each other.

Transparency with Your Congregation

Congregations have a right to know when and how AI is being used in the life of their church. This includes knowing when communications are AI-assisted, how member data is being used, what kinds of personal information are being collected, and what steps are being taken to protect privacy. Churches should know when and how AI is being utilized, what sorts of data are being used, and what steps are being taken to preserve privacy and follow ethical standards.

Transparency builds trust. And trust, once lost, is very difficult to rebuild especially in communities that have historical reasons to be cautious about how institutions handle information about them. Churches can develop trust, deal with problems before they happen, and receive constructive input from members by creating a culture of openness and shared responsibility.

Responsibility in All Uses

Accountability is not a one-time conversation. It is an ongoing practice. Every time AI is used whether for administrative work, content creation, decision-making, or outreach there must be clear rules for how to provide oversight, maintain transparency, and conduct evaluation. Church leaders should establish guidelines regarding who is responsible for reviewing AI outputs, monitoring their impact, and addressing any unforeseen consequences or ethical dilemmas that may arise. Regular evaluations assist in making sure that technology stays in line with the Word of God, the church's mission, values, and the best interests of its members.

One practical structure is a **"Digital Ethics Team"** of three to five people who meet quarterly to review the church's use of digital tools. This group should include at least one pastoral leader and at least one person with expertise in administration or technology. Their role is not to manage every decision, but to ensure that someone is asking the right questions on a regular basis.

Ultimately, accountability ensures that ministry never employs AI without supervision. Instead, it requires constant attention, moral reflection, and a readiness to change or stop using technology when it no longer serves the spiritual health of the community.

Technology Serves the Mission, Not Vice Versa

Church work should always use technology, especially AI, to support the church's spiritual purpose, not as an end in itself. This means that any new digital tool, platform, or AI solution is evaluated and used based on how well it helps advance the Kingdom through worship, discipleship, outreach, and the growth of the faith community.

Leaders and teams in the church need to ask intentionally, "How does this technology help us do what we are called to do?" By tying all their digital choices to the church's principles and vision, they ensure that technology enhances compassion, justice, and service without ever getting in the way of the main work of ministry. This method requires regular reflection, a willingness to change or eliminate technologies that are no longer effective, and the consistent prioritization of people over processes or platforms.

When technology is used to help the church with its mission, it can remain aligned with God's will and meet the changing needs of its community. This protects against mission drift, prevents people from using technology as a measure of success, and reminds everyone that faith, love, and service are always at the heart of ministry.

How AI Retains and Uses Information

When using AI in ministry settings, it is important to understand that these systems often retain the information entered into them and can reference it in future interactions. This means that details provided, such as preferences, interactions, or even sensitive data may be stored and later used to inform recommendations, automate responses, or personalize communications. While this capability can enhance efficiency and user experience, it also raises important questions about privacy, data stewardship, and the ethical use of congregants' information.

Churches must be open about data collection and use, have strong protections in place, and teach their communities how their data is handled. Never assume that information shared with an AI tool disappears after the session ends. Treat every AI interaction involving personal or sensitive information with the same care you would give to a confidential counseling record.

Expanded Reflection: Ethical Guardrails as a Covenant

Ethical guidelines can be presented as a covenant a solemn promise the church makes to its congregation. Such covenants help build trust within the community. A clear and straightforward covenant might include these commitments:

1. We will protect everyone's privacy.
2. We will be open and transparent about our actions.
3. We will actively watch for and address bias.
4. We will ensure people, not just technology, remain responsible.
5. We will pray and seek spiritual guidance before using new tools.
6. We will regularly review and evaluate the impact of our technology use.

To put this covenant into practice, the church should form a small "Digital Ethics Team" of three to five people. This group would meet each quarter to review digital tools, policies, and potential risks. The team should include at least one pastoral leader and at least one member with expertise in administration or technology.

Chapter 9

Faith, Technology, and the Future of the Black Church

We have covered a lot of ground in this book. We have looked at what AI is and how it already shows up in daily life. We have examined the real dangers it poses to privacy, to relationships, to discernment, to the communities most vulnerable to algorithmic bias. And we have explored the genuinely redemptive possibilities: the ways AI can free up time for pastoral care, expand accessibility, support outreach, and strengthen the administrative backbone of ministry.

This book has explored the profound ways artificial intelligence is shaping our world, highlighting both the transformative opportunities and the serious challenges it presents for Christian communities especially within the Black Church. Through scriptural reflection and practical examples, it has become clear that the faithful use of technology is not only possible but essential for advancing God's Kingdom.

As we look to the future, the intersection of faith and technology offers new and dynamic ways to uplift, empower, and serve. With its rich legacy of resilience, advocacy, and spiritual leadership, the Black Church holds a unique position to utilize AI for positive purposes. Faith communities can make sure that technology enhances rather than overshadows the primary goal of ministry by placing a high priority on openness, responsibility, and compassion.

As we look toward the future, the question before the Black Church is not whether to engage with artificial intelligence. That decision has already been made for us, in a sense AI is already here, already shaping the world our congregations inhabit. The question is how we will engage it, and what kind of witness we will offer to the communities watching us navigate this moment. The Black Church has a unique and irreplaceable role to play. With its deep roots in resilience, its long tradition of prophetic leadership, and its hard-won wisdom about navigating systems of power without losing its soul, the Black Church is not just equipped to use AI responsibly it is equipped to model for others what responsible engagement looks like.

The path forward requires intentional practice. Digital literacy must empower all generations. Ministry policies must protect privacy and honor the sacred trust of personal information. Accessibility must ensure that no voice is excluded. Ethical partnerships must be cultivated when technology intersects with counseling, youth engagement, financial stewardship, and community care.

Discernment, prayer, and a commitment to justice must guide the faithful integration of AI. The Black Church's history of turning adversity into opportunity is a testament to the power of hope and innovation rooted in faith. By embracing technology as a tool, never as a replacement for human connection or spiritual wisdom, congregations can navigate change with confidence and grace. In doing so, they strengthen their communities and become a beacon for God's Kingdom in a rapidly evolving world.

A Commission for the Future

As we reach the end of this journey, the enduring message resounds clearly: the Black Church is not merely invited to engage technology it is commissioned to lead within it. In a world reshaped by artificial intelligence and digital innovation, the Church must stand as a beacon of light, guiding its people to resist manipulation, cultivate discernment, and uphold human dignity in every digital space.

Genesis 50:20 reminds us that systems meant for harm are not beyond God's sovereign reach. *"You meant evil against me; but God meant it for good."* This is not passive optimism. It is active theology. It affirms that while human systems may operate with flawed intentions, God's redemptive purposes remain at work often through faithful people willing to steward those systems responsibly.

But beyond policies and procedures lies a deeper calling. The Black Church's legacy of resilience, advocacy, and prophetic leadership uniquely positions it to model what faithful technological stewardship looks like. We have navigated slavery without surrendering hope. We have confronted segregation without abandoning faith. We have transformed suffering into sacred witness. That same theological courage must now guide us through automation and artificial intelligence.

Technology must never become our master. It must remain our instrument. We must refuse the idolatry of innovation while rejecting the fear of progress. We must protect the vulnerable, challenge unjust systems, and insist that every tool we adopt serve the preservation of life and the advancement of justice. This is not merely about adapting to change. It is about shaping it.

Let this be our commissioning:

May the Black Church rise with wisdom equal to the moment. May it steward technology without surrendering theology. May it lead with courage rooted in Scripture and discernment shaped by history.

The God who worked in Egypt still works in the digital age. The God who preserved through empire preserves through algorithms. And as long as the Church remains faithful, what is meant for harm can still be stewarded for good.

Important Warning: Safeguarding Personal and Health Information When Using AI

The use of Artificial Intelligence tools in ministry, administration, counseling preparation, and communication must be approached with heightened responsibility when it involves sensitive information. Church leaders, staff, and volunteers must exercise extreme caution to protect Personally Identifiable Information (PII) and Protected Health Information (PHI).

Personally Identifiable Information (PII) includes data that can identify an individual, such as names, addresses, phone numbers, email addresses, Social Security numbers, financial information, birth dates, or any unique identifying details.

Protected Health Information (PHI) includes any information related to a person's physical or mental health condition, medical history, counseling details, diagnoses, treatment plans, prescriptions, or health insurance data.

AI systems may process, store, or retain input data depending on platform policies. Entering confidential congregational data, counseling notes, financial records, or medical information into AI systems without explicit legal safeguards and secure compliance protocols can expose individuals to privacy breaches, identity theft, discrimination, or reputational harm.

Therefore:

- ☐ Never enter confidential counseling conversations into AI systems.
- ☐ Never upload member databases containing personal details.
- ☐ Never input medical, mental health, or financial information tied to identifiable individuals.
- ☐ Never use AI platforms as substitutes for secure record-keeping systems.
- ☐ Always review platform privacy policies and data retention terms before use.
- ☐ Ensure compliance with applicable federal and state privacy laws, including HIPAA where health-related information is involved.
- ☐ Obtain informed consent where digital tools are used in communication or administrative processes.

The sacred trust between shepherd and congregation requires vigilant protection of personal information. Privacy is not merely a legal obligation; it is a pastoral responsibility and a moral covenant. Technology must never compromise the dignity, confidentiality, or safety of God's people.

If the information is confidential in the sanctuary, it must remain confidential in the digital space.

About the Author

Rev. **Dr. Sineta Scott Robertson** is a theologian, author, and executive pastor devoted to empowering the Black Church for effective leadership in an ever-evolving world. She currently holds a leadership role at **Community Bible Baptist Church**, where her ministry emphasizes preaching, teaching, discipleship, and strategic development. Her pastoral approach centers on upholding biblical integrity, fostering cultural awareness, and equipping leaders to serve wisely and courageously.

Dr. Robertson also serves as the **Chief of Staff** and faculty of the **Washington Baptist Theological Seminary**, where she invests in the next generation of ministers and church leaders. In both the pulpit and the classroom, she challenges students and congregations alike to think critically, engage Scripture deeply, and respond to contemporary issues through a sound theological lens.

She has authored *AI and the Black Church: Bridging Faith, Justice and Innovation* to compassionately reach her peers and educate them on Artificial Intelligence. She previously wrote ***The Promise of the Peace of God: Waiting for God During a Pandemic***, reflecting on God's promises during the COVID crisis; one of America's darkest periods. Her personal experiences of God's peace amid loss shape her commitment to faith and tranquility during difficult times. Through her writing, Dr. Robertson connects theology with real-life situations, supporting churches as they stay grounded in tradition while facing contemporary issues.

With over twenty years of ministry experience and a Doctorate in Christian Counseling, Dr. Robertson is recognized for her prophetic insight, strategic leadership, and steadfast dedication to preparing the Church for the future. Her work blends faith, innovation, leadership, and justice, consistently rooted in Scripture and prayer.

She is married to Deacon Vernon Robertson and is the mother of Rev. Dr. Angelo Sullivan and Jermeka Scott.

Bibliography

The following resources informed the research, theological grounding, and practical guidance presented throughout this book. Readers are encouraged to explore these sources for deeper study on the intersection of faith, ethics, and technology.

Scripture

The Holy Bible, English Standard Version (ESV). Wheaton, IL: Crossway, 2001.

The Holy Bible, King James Version (KJV). Cambridge: Cambridge University Press, 1611.

The Holy Bible, New International Version (NIV). Grand Rapids, MI: Zondervan, 2011.

Theology and Biblical Studies

Cone, James H. *The Cross and the Lynching Tree.* Maryknoll, NY: Orbis Books, 2011.

Cone, James H. *A Black Theology of Liberation.* 40th anniversary ed. Maryknoll, NY: Orbis Books, 2010.

Frazier, E. Franklin, and C. Eric Lincoln. *The Negro Church in America / The Black Church Since Frazier.* New York: Schocken Books, 1974.

Grudem, Wayne. *Systematic Theology: An Introduction to Biblical Doctrine.* 2nd ed. Grand Rapids, MI: Zondervan Academic, 2020.

Lincoln, C. Eric, and Lawrence H. Mamiya. *The Black Church in the African American Experience.* Durham, NC: Duke University Press, 1990.

West, Cornel. *Prophesy Deliverance! An Afro-American Revolutionary Christianity.* Philadelphia: Westminster Press, 1982.

Artificial Intelligence and Technology

Autor, David. "Work of the Past, Work of the Future." *AEA Papers and Proceedings* 109 (2019): 1-32.

Buolamwini, Joy, and Timnit Gebru. "Gender Shades: Intersectional Accuracy Disparities in Commercial Gender Classification." *Proceedings of Machine Learning Research* 81 (2018): 1-15.

Crawford, Kate. *Atlas of AI: Power, Politics, and the Planetary Costs of Artificial Intelligence.* New Haven, CT: Yale University Press, 2021.

Eubanks, Virginia. *Automating Inequality: How High-Tech Tools Profile, Police, and Punish the Poor.* New York: St. Martin's Press, 2018.

Floridi, Luciano, ed. *The Ethics of Artificial Intelligence.* Cambridge, MA: MIT Press, 2023.

Metz, Cade. *Genius Makers: The Mavericks Who Brought AI to Google, Facebook, and the World.* New York: Dutton, 2021.

Noble, Safiya Umoja. *Algorithms of Oppression: How Search Engines Reinforce Racism.* New York: NYU Press, 2018.

Russell, Stuart. *Human Compatible: Artificial Intelligence and the Problem of Control.* New York: Viking, 2019.

AI Ethics and Policy

Floridi, Luciano, et al. "An Ethical Framework for a Good AI Society: Opportunities, Risks, Principles, and Recommendations." *Minds and Machines* 28, no. 4 (2018): 689-707.

IEEE Global Initiative on Ethics of Autonomous and Intelligent Systems. *Ethically Aligned Design: A Vision for Prioritizing Human Well-being with Autonomous and Intelligent Systems.* 1st ed. New York: IEEE, 2019.

Partnership on AI. *About ML: Annotation and Benchmarking on Understanding and Fairness of Machine Learning Systems.* New York: Partnership on AI, 2019.

United States Department of Health and Human Services. *Health Insurance Portability and Accountability Act of 1996 (HIPAA).* Washington, DC: U.S. Government Publishing Office, 1996. https://www.hhs.gov/hipaa.

Black Church History and Civil Rights

Branch, Taylor. *Parting the Waters: America in the King Years, 1954-63.* New York: Simon & Schuster, 1988.

Du Bois, W. E. B. *The Souls of Black Folk.* Chicago: A. C. McClurg & Co., 1903. Reprint, New York: Dover, 1994.

Giddings, Paula J. *Ida: A Sword Among Lions.* New York: Amistad, 2008.

King, Martin Luther, Jr. *Strength to Love.* New York: Harper & Row, 1963. Reprint, Minneapolis: Fortress Press, 2010.

Raboteau, Albert J. *Slave Religion: The "Invisible Institution" in the Antebellum South.* Updated ed. New York: Oxford University Press, 2004.

Faith and Technology

Estes, Douglas. *SimChurch: Being the Church in the Virtual World.* Grand Rapids, MI: Zondervan, 2009.

Hipps, Shane. *The Hidden Power of Electronic Culture: How Media Shapes Faith, the Gospel, and Church.* Grand Rapids, MI: Zondervan, 2006.

Schuurman, Derek C. *Shaping a Digital World: Faith, Culture and Computer Technology.* Downers Grove, IL: IVP Academic, 2013.

Techner, Glen. *Digital Discipleship: Growing Your Faith in a Screen-Saturated World.* Nashville: Abingdon Press, 2021.

Additional Resources for Further Study

AI Now Institute. *Annual Report.* New York: AI Now Institute, New York University. https://ainowinstitute.org.

Berkman Klein Center for Internet & Society. "AI Ethics Resources." Harvard University. https://cyber.harvard.edu.

Data & Society Research Institute. *Reports and Working Papers.* New York: Data & Society. https://datasociety.net.

Pew Research Center. "Public Attitudes Toward AI." Washington, DC: Pew Research Center. https://pewresearch.org.

Made in the USA
Monee, IL
04 November 2026

67381278R00046